AF599013

ANIMALS AROUND THE WORLD
ALL ABOUT
AUSTRALIAN
TASMANIAN
DEVILS
EZ
READERS
Tracy Vonder Brink

Creating Young Nonfiction Readers

EZ Readers lets children delve into nonfiction at beginning reading levels. Young readers are introduced to new concepts, facts, ideas, and vocabulary.

Tips for Reading Nonfiction with Beginning Readers

Talk about Nonfiction

Begin by explaining that nonfiction books give us information that is true. The book will be organized around a specific topic or idea, and we may learn new facts through reading.

Look at the Parts

Most nonfiction books have helpful features. Our *EZ Readers* include color photographs and graphic aids, a table of contents, a glossary, and an index. Share the purpose of these features with your reader.

Color Photos and Graphic Aids

A lot of information can be found by "reading" photos, charts, maps, and other graphic aids found within nonfiction texts. Help your reader learn more about the different ways information can be displayed.

Table of Contents

Located at the front of the book, this list shows the big ideas within the text and the page numbers where they can be found.

Glossary

Located at the back of the book, the glossary defines key words and phrases that are related to the topic. These words and phrases can be found in the text in colored type.

Index

Located at the back of the book, an index is an alphabetical list of topics and the page numbers where they can be found.

With a little help and guidance about reading nonfiction, you can feel good about introducing a young reader to the world of *EZ Readers* nonfiction books.

Mitchell Lane
PUBLISHERS

2001 SW 31st Avenue
Hallandale, FL 33009
www.mitchelllane.com

First Edition, 2025.

Author: Tracy Vonder Brink
Designer: Rhea Magaro
Editor: Kim Thompson

Names/credits:
Title: All About Australian Tasmanian Devils / by Tracy Vonder Brink
Description: Hallandale, FL :
Mitchell Lane Publishers, [2025]

Series: Animals Around the World
Library bound ISBN: 978-1-68020-950-1
eBook ISBN: 978-1-68020-968-6

EZ Readers is an imprint of
Mitchell Lane Publishers

PHOTO CREDITS
Alamy: Dave Watts, 16-17, 18-19; iStock: BerndC, cover, 1; BlackAperture, 12-13; JurgaR, 20-21; Shutterstock: Miela197, 4-5; james_stone76, 6-7; Susan Flashman, 8-9; JCMP, 10-11; Martin Mels, 14-15

CONTENTS

Tasmanian devils lived in **mainland** Australia hundreds of years ago. But the **climate** became too dry there. **Dingoes** also hunted them. Today they are only found on the island of Tasmania.

Tasmanian devils have black or dark brown fur. Many have patches of white on their chests or bottoms. They have fluffy tails. Their round ears are pink inside. They are approximately the size of a small dog.

Tasmanian devils are noisy. They growl, bark, and screech. They make loud sounds to chase away other Tasmanian devils.

Tasmanian devils live alone. They spend the day sleeping in a hollow log, under a bush, or in a pile of rocks. They come out at night to hunt.

Tasmanian devils have sharp teeth. They hunt rabbits, snakes, and other small animals. They see well in the dark. Their ears help them find **prey**. Their noses sniff out food.

Tasmanian devils are also **scavengers**. They eat almost any dead animal they can find. Their strong jaws and teeth can crunch up bones.

Tasmanian devils are **endangered** because of a sickness. It makes lumps grow on their faces and in their mouths. They die because they can no longer eat. Scientists are working on ways to help them.

Like other Australian animals, Tasmanian devils are **marsupials**. That means mothers raise their babies in a pouch. A baby Tasmanian devil is called a joey. A mother's pouch has room for four joeys.

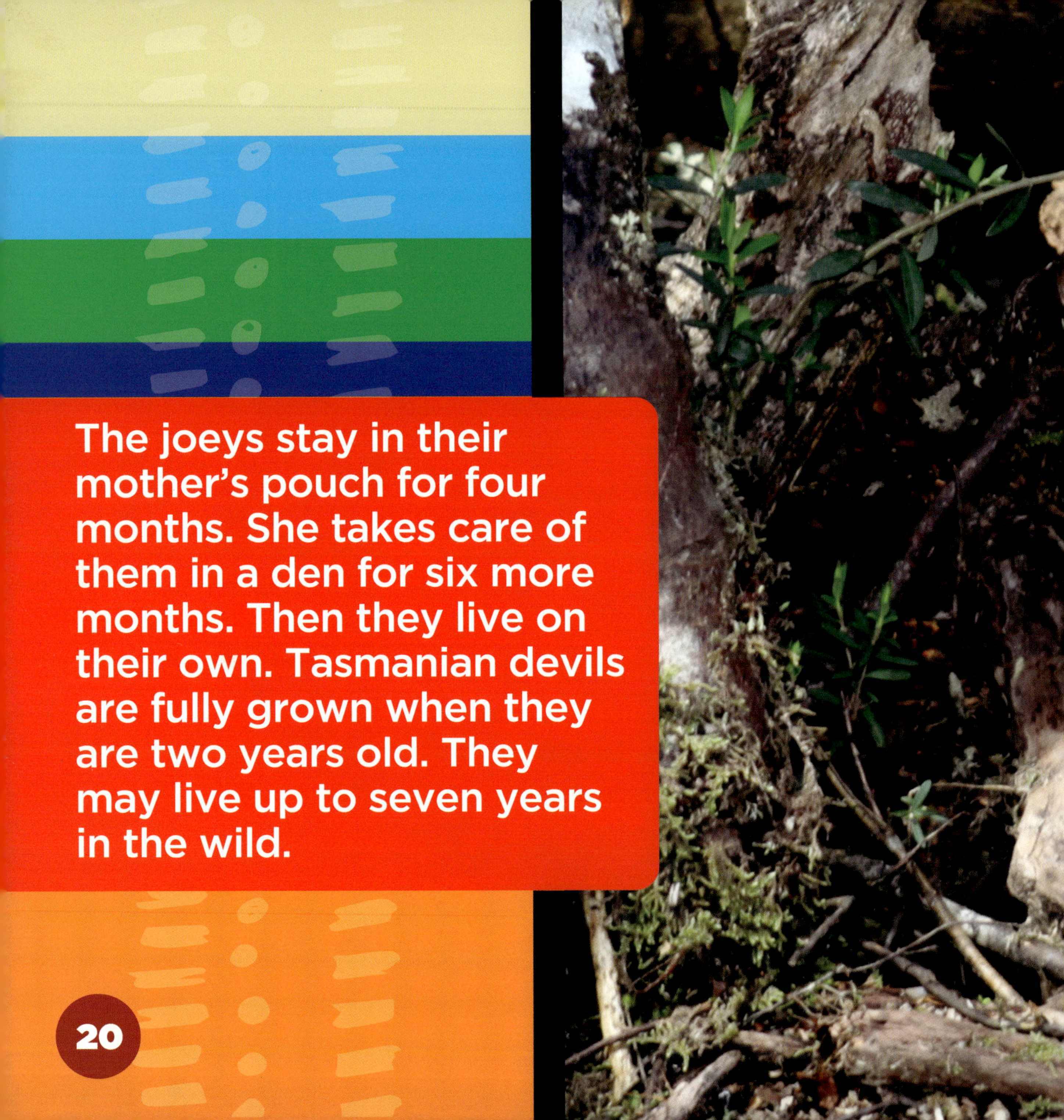

The joeys stay in their mother's pouch for four months. She takes care of them in a den for six more months. Then they live on their own. Tasmanian devils are fully grown when they are two years old. They may live up to seven years in the wild.

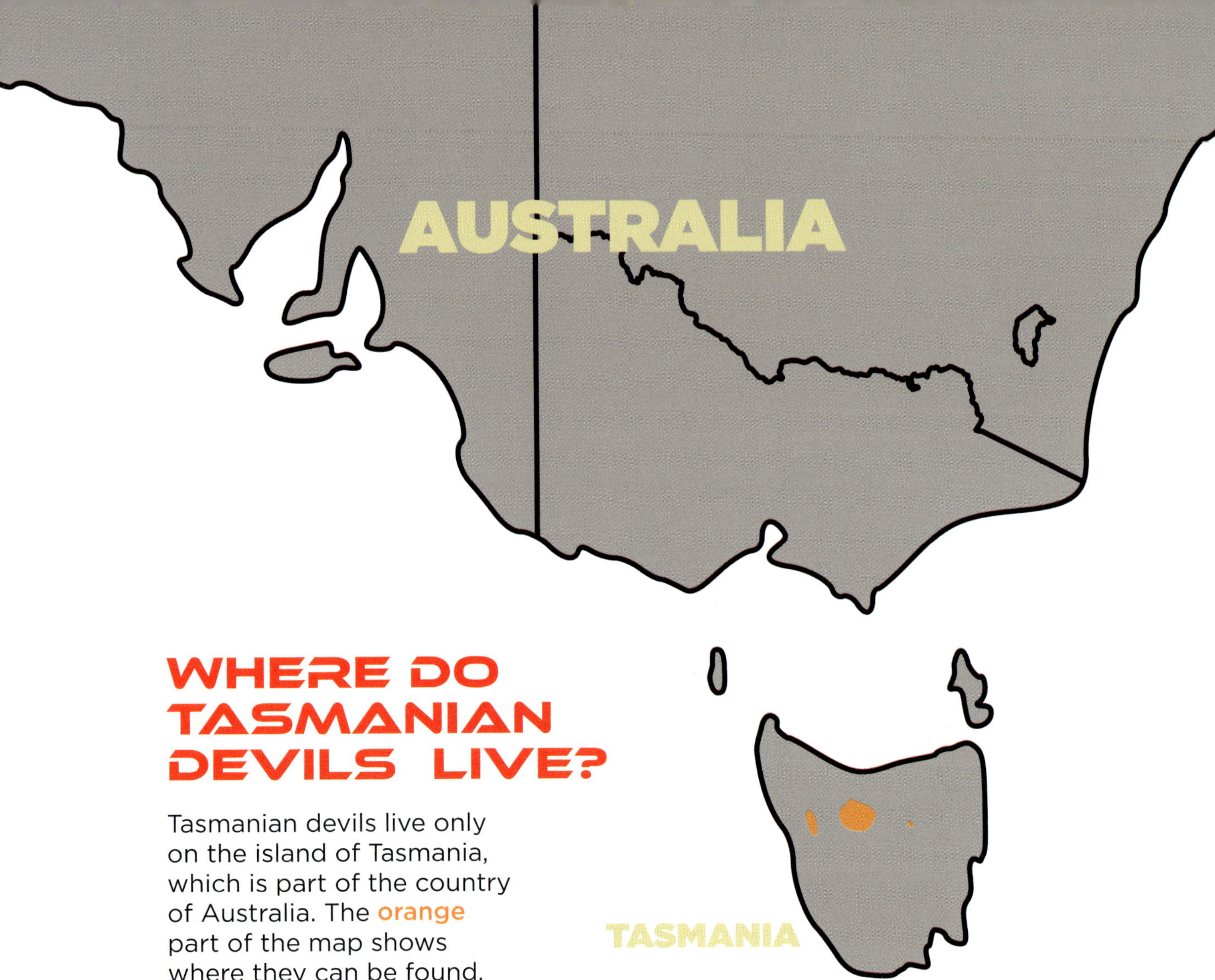

WHERE DO TASMANIAN DEVILS LIVE?

Tasmanian devils live only on the island of Tasmania, which is part of the country of Australia. The orange part of the map shows where they can be found.

INTERESTING FACTS

- Like skunks, Tasmanian devils can make a bad smell when they are upset.
- Tasmanian devils may travel up to nine miles (14 kilometers) every night to find food.
- Tasmanian devils sometimes like to lie out and sun themselves.
- Tasmanian devils can run at least as fast as a person can.
- Tasmanian devil joeys are also called imps.

PARTS OF A TASMANIAN DEVIL

Ears
The pink parts of their ears turn red when Tasmanian devils are upset or excited.

Jaws
A Tasmanian devil's strong jaws let it bite as hard as a dog that is four times its size.

Legs
A Tasmanian devil's front legs are longer than its back legs.

Tail
Healthy Tasmanian devils have fat tails.

Whiskers
A Tasmanian devil's long whiskers help it feel around in the dark for food.

GLOSSARY

climate (KLYE-mit)
The usual weather of an area over a long period of time

dingoes (DIN-gohs)
Australian animals related to wild dogs

endangered (en-DAYN-jurd)
In danger of dying out, often because of human activity

mainland (MAYN-luhnd)
The largest or main part of a country that does not include its islands

marsupials (mahr-SOO-pee-uhlz)
Animals that carry their babies in a pouch on the front of the female's body

prey (pray)
Animals that are hunted by other animals for food

scavengers (SKAV-uhn-jurz)
Animals that will eat dead or rotting food

INDEX

FURTHER READING

Grack, Rachel. *Tasmanian Devils*. Minneapolis, MN: Bellwether Media, 2023.

Markle, Sandra. *Tasmanian Devils: Nature's Cleanup Crew*. Minneapolis, MN: Lerner Publications, 2024.

ON THE INTERNET

San Diego Zoo: Tasmanian Devil
https://animals.sandiegozoo.org/animals/tasmanian-devil
Read about Tasmanian devils from a zoo that takes care of them.

Tasmanian Department of Resources and Environment: Tasmanian Devil Facts for Kids
https://nre.tas.gov.au/conservation/threatened-species-and-communities/lists-of-threatened-species/threatened-species-vertebrates/save-the-tasmanian-devil-program/tasmanian-devil-information-for-kids/tasmanian-devil-facts-for-kids
Learn lots about Tasmanian devils from this Tasmanian government site.